Bibliographic information published by the German National Library:

The German National Library lists this publication in the National Bibliography; detailed bibliographic data are available on the Internet at http://dnb.dnb.de .

Imprint:

Print and binding: Books on Demand GmbH, Norderstedt Germany
ISBN: 9783668442870

This book at GRIN:

http://www.grin.com/en/e-book/359055/about-food-justice-and-food-sovereignty-industrial-versus-natural-food

Aya Tomozawa

About food justice and food sovereignty. Industrial versus natural food

GRIN Publishing

Aya Tomozawa

8 May 2017

Food Justice Research Paper

Industrial food producing corporations provide consumers little to no options to choose healthy and environmentally friendly food. This lack of choices violates people's food sovereignty, or rights to easily access nutritious, safe, and local food. Industrially produced food is packed with high levels of preservatives, sugar, sodium, and synthetic chemicals; however there are alternatives such as local farmer's markets, but it seems that they only attract a niche category of pretentious hipsters as the customer base. Upon my visits to the local farmer's market, I discovered that these stereotypes of exclusivity were not accurate and that farmer's markets had a surprisingly large variety of options that I could choose from. Through research from Michael Pollen's *The Omnivore's Dilemma*, Raj Patel's *Stuffed and Starved,* Robert Gottlieb and Anupama Joshi's *Food Justice*, as well as my observations at the Broad Ripple Farmer's Market, I found that the issues of inadequate food sovereignty that plague human and environmental health go beyond the lack of choices for consumers. Food sovereignty violations in today's society are due, but not limited to, the stronger relationships the state has with the market over civil society, lack of accurate education about food production and costs that ensue, and through the difficulties that exist for consumers to access to fresh locally produced food.

I decided to visit the Broad Ripple Farmer's Market once in September and the second time in October as my field research. The first time I attended I was overwhelmed by the numbers of people, stands, and produce that I had never seen before. There were stands selling fresh fruit and vegetables, pasture fed meat, baked goods, coffee and tea beverages, cooking utensils, and even naturally produced dog treats. My misconception that farmer's markets products were of a limited variety could not have been more wrong. Eventually I got the courage to go up to a stand, ask a couple questions about their production process, and purchase some food to test whether Michael Pollan was correct in that the taste of pastoral food was distinctively more flavorful. In addition to inquiring about products that I did not recognize, I ended up purchasing apple butter at Wildflower Ridge Honey, a Jonathan apple from Wild's Apple Farm, lettuce from Funny Bone Farm, chicken from Simpson's Farm, and eggs from Phelp's Family Farm.

One of my first experiments was with chicken from Simpson's Farm. The man who sold me the chicken was named Darby Simpson and he explained to me that his farming methods were was basically the same as Joel Salatin, the pastoral farmer Pollan shadowed in his book, but on a much smaller scale. His family farm had been passed down from generation to generation, but had taken a financial hit from the pressures of industrial farming companies. He told me that about a decade ago he and his wife moved back to Martinsville and began raising grass-fed livestock without the use of antibiotics. Similarly to Salatin's explanation to Pollan, Mr. Simpson informed myself that the extra care he put into raising his farm animals did not require him to use chemical medications that are present on industrial farms. He told me that at first people had been hesitant to buy from him because of his refusal to use vaccinations or antibiotics on his livestock. However Mr. Simpson had done his research before taking over the farm and promised me that they were probably safer than the meat I would buy at the grocery store. Before my conversations with Mr. Simpson, I never before had such an in-depth conversation about how and why my food was made. Since I had read *The Omnivore's Dilemma* already I found that my interactions with Mr. Simpson reminded me so much of Joel Salatin's account with Pollan because both of these pastoral farmers seemed very open to explaining all the biology that is required to know in order to properly running their ecologically efficient farms.

I ended up buying three chicken legs from Mr. Simpson and pan grilled them with barbeque sauce. After cooking the chicken I ate my finished product and it really was tastier than any other barbequed chicken I had ever eaten before. While I didn't use natural ingredients like Pollan did in the chapter, *The Meal: Grass Fed,* I definitely noticed a distinct difference in the flavor of Simpson's chicken legs and the ones that I usually buy from Marsh. Now I am a person whose cooking skills are mediocre at best, so the only conclusion that I can find to be possible is that the quality of Mr. Simpson's chicken must have been superior. The second time I bought ground chicken from the Mr. Simpson and made chicken meatballs. Once again the end product was delicious and I think that at least in terms of meat, Pollen wasn't lying that pastoral farm was better in terms of taste than the industrially produced counterparts.

Taste might not seem like an important component to food sovereignty, however in Raj Patel's text, *Stuffed and Starved*, his first point of his outline on changing the current unhealthy and unsustainable food system is to transform consumer tastes (309). I would consider myself to be part of the majority who grew up on industrially produced food and therefore my physiological desires have been shaped to crave processed foods that had come from industrial food companies. The logic from Patel's point is that people will be more willing to purchase better quality food if they have actually had the edible experience. An obstacle that exists to change just one person's taste preferences will take time and consistent consumption of pastoral produce in order for any transformations to occur. I can relate to this because while I would like to eat Simpson Farm meat again, I still find myself craving synthetic chemical rich foods like Lucky Charms. However consumer "needs" are only a miniscule factor that requires change in order to improve upon the ecological and human health conditions as well as the social inequalities that still exist surrounding food.

The prices of food from the farmers market was indeed more expensive than at the grocery stores where I usually buy food. For example the eggs that I bought from Phelp's Family Farm cost me $4.00 for a dozen eggs where at Marsh I can buy the same number for less than half the price. This was because Phelp's Family Farm, as well as other local farmers, takes into account the costs of production, packaging, and transportation. There are not hidden costs which is not transparent from industrially produced food hides the total costs through cheap prices. Being a college student living on a tight budget, I can see why it would not be desirable for other low income people to pay the additional amount just for better quality. However before reading Pollan and Patel's books, I did not realize that the higher costs would actually be lower in the long run. The explanation Pollan received in regards to the price differences resulted with Joel Salatin stating, "Society is not bearing the cost of water pollution of antibiotic resistance, of food-borne illnesses, of crop subsidies, of subsidized oil and water---of all the hidden costs to the environment and the taxpayer that make food seem cheap," (243). This means that while monetarily I am paying less for the grocery store eggs, I am actually being inflicted with excess hidden costs that come from the factors that Salatin mentioned above. These excess hidden costs from industrial food companies include the depletion of fresh drinking water, destruction of fertile land, and harmful chemicals that animals and people consume due to the use of substances such as pesticides.

These additional costs when taken into account result in a price much greater than the $4.00 pastorally grown eggs. This actual lower price is because pastoral farms like Phelp's Family Farm do not produce excess waste or pollution with their sustainable farming methods. The true cost was completely represented in the $4.00 price of Phelp's eggs where deceit of the true cost continues to exist with the industrial products sold at chain grocery stores. In terms of food sovereignty this practice of industrial food companies masking the actual costs of their products by advertising lower prices should not be considered as ethical nor fair for any consumer, especially for people with lower incomes. The lack of transparency in the true costs of industrialized food violates peoples' rights as consumers to have the abilities to choose the healthier and more cost-effective options. While this behavior from industrial food corporations is unethical, it remains in practice because of the state allowing the market to continue to produce and sell their products in this way. The state has continued their lack of regulation with the misconception that the market behaves in a naturally beneficial manner. However there is nothing natural about the market and the behavior only benefits the industry at the civil society sector's expense.

While deception in prices continue to trick consumers, industrially produced food also has not allowed people to accurately choose products that are beneficial for their home-grown businesses and natural environments. Multi-billion dollar industries like Walmart have used bullying tactics to drive out local competition in areas ranging from small towns to urban cities with little to no regard to the local environment. For example in places where Walmart-like groceries failed to meet the corporate standards, they would just abandon the lot leaving only destroyed land (Gottlieb, Joshi 157-159). With the removal of the many smaller stores, less possible choices for consumers will be available to purchase food. Also if the industrial stores went out of business, the people of the area would then have nowhere locally to purchase food since all of the smaller businesses had been pushed out. Ironically industrial grocery stores tend to either establish their locations away from lower income neighborhoods, creating food deserts or as Patel phrased it, "Areas in which it is extremely hard to access fresh food without a car." (Patel 249). The discriminatory strategies used by big industrial corporations represents the apathy the market has towards the well-being of their consumers and the poorer sectors of society they tend to avoid.

Farmer's markets do allow consumers to make better ecological decisions when purchasing naturally grown food. However farmer's markets have tended to not be prevalent in lower-income areas. I looked up the places where farmer's markets are located and most of them are north of downtown Indianapolis which is not where the majority of low-income people and food deserts are located (United States Department of Agriculture Economic Research Service). While the intentions of farmer's markets are positive in that they provide healthier and environmentally friendly choices, their lack of presence in areas that have little to no food sovereignty I find to be perplexing. The Broad Ripple Farmer's Market did cater to lower-income people by accepting food stamps and income supplementing coupons for payment, but the fact that they are not walking distance away from downtown Indianapolis and are only in active once a week for three to four hours does not really benefit nor attract the lower income demographic to purchase farmer's market produce.

Problems with food sovereignty don't just exist at local levels in the Indianapolis. All the way up through the national government, food sovereignty has been limited. This is a world-wide issue that has much to do with the neo-liberal favoring the interests of markets over the rights of consumers and laborers. According to Dr. Van Gerven, there should be a balance between the states interest with the market and civil society. However organizations such as the United States Department of Agriculture (USDA) have continuously passed series of Farm Bills that almost exclusively benefits the market sector. Inside the market sector are the industrial food corporations who have squandered the number of options consumers are able to make in regards to food. While the farmer's markets are part of the market sector, they generally have little to no political clout when it comes to the USDA's decisions to pass Farm Bills (Gottlieb, Joshi 79-83). Unlike industrial food corporations, farmer's markets care more about the quality of their products instead of the profits. Industrially produced food companies care more about profits and quantity of products sold. However money has been a frequent characteristic in history of buying influence in all levels of government.

Even though government and industrial conglomerates have taken over most of the food market and prevented consumers from their freedom to choose better options of food, there is still hope that alternative food producers like farmer's markets can help improve upon this issue. Educational programs in regards to environmentally beneficial farming have begun to emerge in areas around the world. For example the Landless Rural Movement (MST) that began started in Brazil during the 1960s brought displaced people together to revive abandoned land and create small sustainable communities. The incredible success of this movement has been attributed to their focus on educating the people about how to grow food pastorally and where the food actually comes from (Patel 211-213). I believe that education is the key for people to realize that their food sovereignty has been violated because before taking this class I thought that I had the independence to choose whatever I wanted to eat. The Broad Ripple Farmer's market had aspects of the MST education through the conversations I had with Mr. Simpson and other participants in the farmer's market. Through reading about the MST, TI-261 class, and my trips to the farmer's market I believe more than ever that there needs to be a greater emphasis on educating people about the inequalities from the industrial farming system and how we as consumers can work together to combat this world-wide issue.

My experience attending and consuming farmer's market produce has helped me understand why non-industrial food production is necessary to combat suppressed food sovereignty. While the quantity of products are less than at the industrially produced grocery counterparts, there was still more variety of options the farmer's market. Instead of choosing from different forms of corn-based products produced from industrial corporations, the farmers market had food items that I didn't even know existed. Food sovereignty should be something that all people have the right to express, but this cannot happen unless people realize that it is an issue.

Works Cited

Gottlieb, R. and Joshi, A. (2010). *Food Justice*. Massachusetts: The MIT Press.

Patel, R. (2007). *Stuffed and Starved: The Hidden Battle for the World Food System*. New York: Melville House.

Pollan, M. (2006). *The Omnivore's Dilemma: A Natural History of Four Meals*. Penguin Press.

United States Department of Agriculture Economic Research Service. (19 October 2016). *USDA Economic Research Service Atlas*. Retrieved from http://www.ers.usda.gov/data-products/food-access-research-atlas/go-to-the-atlas.aspx.